Siebel Fh 104 V 2 D-IMCH. Unlike on the V 1 the landing gear here has already been modified to the single strut types.

Siebel Fh 104/Si 204

and License Variants

by Harald Vogt

Developmental History

The Siebel Si 204's developmental history begins with the designing of the well-known Siebel Fh 104 Halore.

This five-seat, twin-engine passenger aircraft was originally conceived by the firm of Klemm in Böblingen as the Klemm Kl 104.

Dipl.Ing. Friedrich Fecher bore overall responsibility for the design.

When a Klemm branch was established in Halle, Fecher was transferred to the Flugzeugbau Halle GmbH (as it became known) and took all his design blueprints and documents with him to Halle, where construction on the design continued.

As a result of the Halle facility becoming an independent operation the aircraft's designation was changed to Fh 104.

In 1936 the famous sport flyer Friedrich Wilhelm Siebel assumed the directorship of the company. In doing so, the company's title was changed to the Siebel Flugzeugwerke Halle GmbH.

The Siebel Fh 104 Halore's (D-IQPG) maiden flight took place on 25 February 1937. Production began the same year.

The design proved to be both successful and popular, becoming world famous after completing many record-setting and competition flights.

1938: Victor in the III Raduno del Littorio.
1938: One-day European circuit via 12 countries over 6200 km in 21 flying hours.
1938: Fastest aircraft at the 1st International Air Race in Frankfurt.
1939: Fastest aircraft at the 2nd International Air Race in Frankfurt.
1939: Africa flight of over 40,000 km.

Design Features

Type: Twin-engined passenger aircraft
Wings: Twin-spar cantilever, plywood covered with landing flaps.
Fuselage: All-metal monococque with oval cross-section.
Empennage: All-metal cantilever design with trim tabs on rudder and elevators.
Landing gear: Hydraulically retractable with forked struts. Housed in engine gondolas. Fixed tailwheel and hydraulic brakes. The struts were straight on the A-series.
Engine: 2 x Hirth HM 508C engines, each with 240 hp.
Production version was fitted with 2 x Hirth HM 508D 280 hp engines.
Several changes were made during the testing phase.
The landing gear was altered to a single strut vice the forked design. It was also possible to fit the landing gear with snow skis.

Siebel Fh 104 A

First production variant, which entered production in 1939 with an initial batch of 11 aircraft.

Production continued until 1942 and ended following a total run of 46 aircraft.

A follow-on development, designated the Siebel Fh 104 C with Argus AS 10C 240 hp engines, never saw production.

Siebel Si 204 Developmental History

In 1939/40 the first blueprints were drawn up for a successor to the popular Fh 104 which took on the form of an eight seat liaison airplane.

Construction was completed in a relatively short time and on the 23rd of May, 1941 the prototype - as the Siebel Si 204 V 1 - had completed its first successful maiden flight.

The aircraft was powered by two Argus AS 410A, each with an output of 465 hp.

Many features from the Fh 104 were carried over into the new design. However, it should be mentioned that the Siebel Si 204 was a completely new aircraft.

There were significant differences in the following areas.

The wings were of an all-metal construction. Designers opted for twin rudders in place of the single rudder design of the Fh 104.

The documented results of successful flight trials carried out by the two prototypes led to the production of a small batch of Siebel Si 204 A-0/A-1s.

From early on the Luftwaffe showed interest in the aircraft design, which was well-suited to the training and flying classroom roles, plus other diverse tasks.

The first prototype which was based on the Luftwaffe's requirements and which evolved into the Siebel Si 204 D was the Siebel Si 204 V 3.

Cockpit view of the Siebel Fh 104.

Below: Military version of the Siebel Fh 104 (military designation NB+AS) during testing with snow skids.

The most obvious feature of the D-model was the fully glazed nose section, which proved particularly suited for the primary training of the Luftwaffe's bomber crews.

The similarity of the instrument panel layout to that found in bombers was also a significant advantage in providing bomber crews with familiarization training.

Plane for plane, the Siebel Si 204 D replaced the Focke-Wulf Fw 58 Weihe in the instrument flight training role; the latter aircraft, with its fabric covered design, displayed critical disadvantages in extreme weather conditions compared with the all-metal construction of the Siebel.

The Siebel Si 204 D also saw use as a liaison aircraft in nearly all theaters.

Since the Siebel Flugzeugwerke Halle GmbH was heavily involved in building bombers for the Luftwaffe (Ju 88/Ju 188), production of the Siebel Si 204 D was transferred to facilities in occupied France and in Czechoslovakia.

Production in France got underway relatively slowly, whereas in Czechoslovakia - then known as the Protectorate of Bohemia and Moravia - production began without any difficulties.

Follow-on developmental programs and documentation remained as before with the Siebel Flugzeugwerke in Halle.

Siebel Si 204 S early production model. Note the type designation below the cockpit. This aircraft was probably used for promotional flights. The exact aircraft coding isn't known.

An armed bomber variant, designated the Siebel Si 204 E, plus a radar-equipped version for training night fighter crews on the use of this new technology were also developed.

In addition to its many other roles the aircraft was also used for training Me 262 crews on single engine handling characteristics.

Due to the general shortage of material, design plans were drawn up for converting production of the wings and control surfaces to wood.

However, it is unknown whether this partial conversion to wood actually ever materialized.

A Siebel Si 204 A (D-ASGU) in civilian use.

Siebel Si 204 A (RE+EY) in Luftwaffe service. The Si 204 A could also be fitted with skis.

This Siebel Si 204 A belonged to Fluzeugführerschule FFS (B) 20 at Rosenborn.

Siebel Si 204 As were popular aircraft at B and C type training units. This is a Siebel Si 204 A of the Flugzeugführerschule FFS (C11) in Zeltweg, Austria.

Siebel Si 204 A of an unknown Luftwaffe unit. The camouflage indicates this may have been a liaison aircraft with a night fighter unit. This aircraft has been mentioned in aviation literature in connection with testing of the Lippisch DM-1 in Prien. However, further details are not known.

Siebel Si 204 D in Luftwaffe Service

Scrapbook photo taken in front of a Siebel Si 204 D-1 in Prague-Ruzine.

A Siebel Si 204 D of Flugzeugführerschule (B) 9. The photo was taken in 1944.

On 7 May 1945 this aircraft landed in Belp, south of Bern (Switzerland) with five people on board. The military code DL+NT can be seen clearly here. National markings were painted out by the crew prior to taking off. According to eyewitness reports the paint was still fresh when the aircraft landed.
Subsequent official reports listed the aircraft as being in a run-down condition.
Radio equipment consisted of FuG 10, FBl 2 and EBl 3 sets.
The airplane was flown to Dübendorf on May 9th, 1945, where this photo was taken.
Following a general overhaul the aircraft was absorbed into the Swiss Flugwaffe inventory under the designation B-3.

Five highly motivated pilot trainees in front of a Siebel Si 204 D-1 from Flugzeugführerschule FFS (B) 11. At the time this photo was taken (1941) training operations were being conducted at Aigen Enstal.

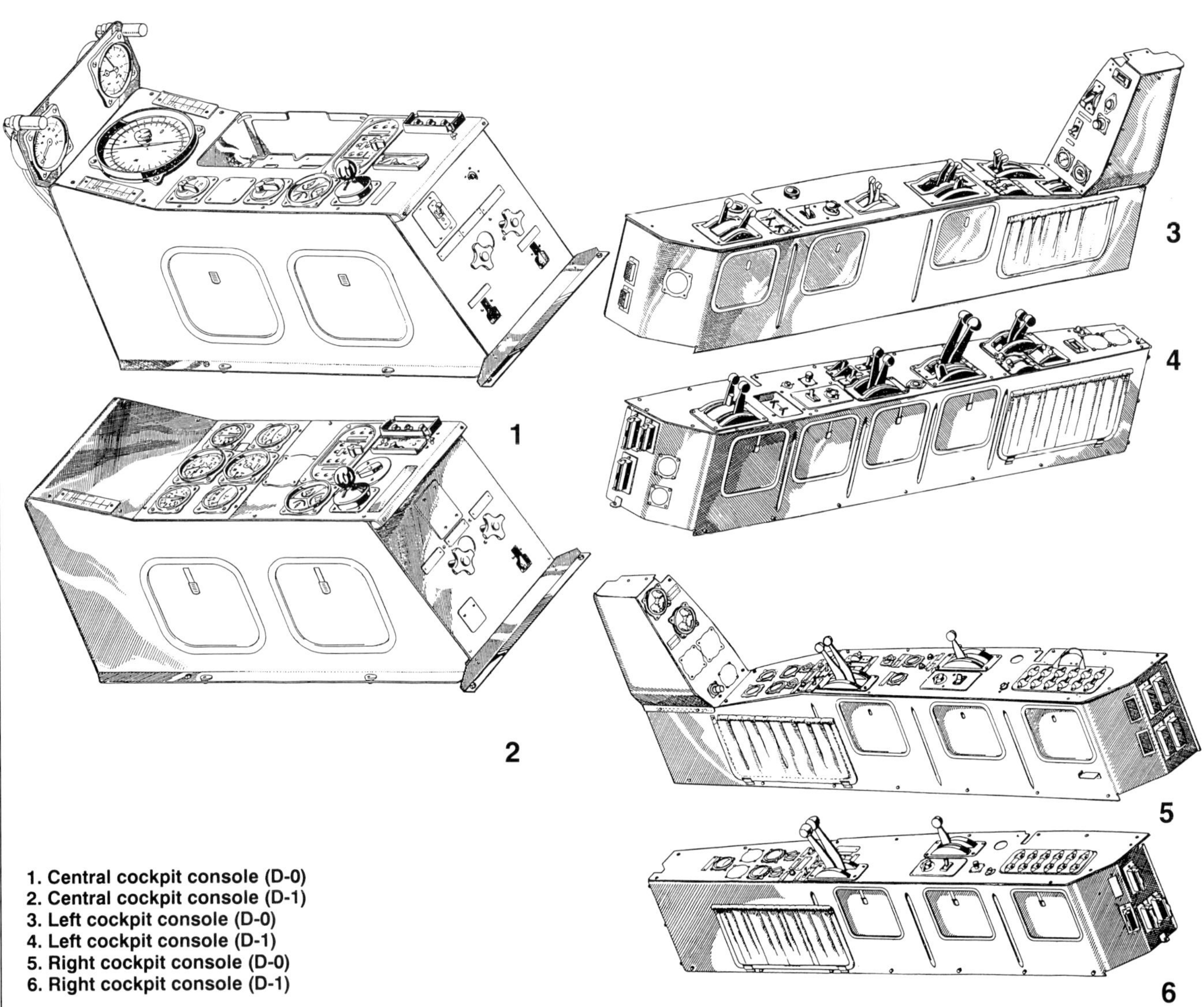

1. Central cockpit console (D-0)
2. Central cockpit console (D-1)
3. Left cockpit console (D-0)
4. Left cockpit console (D-1)
5. Right cockpit console (D-0)
6. Right cockpit console (D-1)

The fully-glazed cockpit of the D-series offered an excellent view.

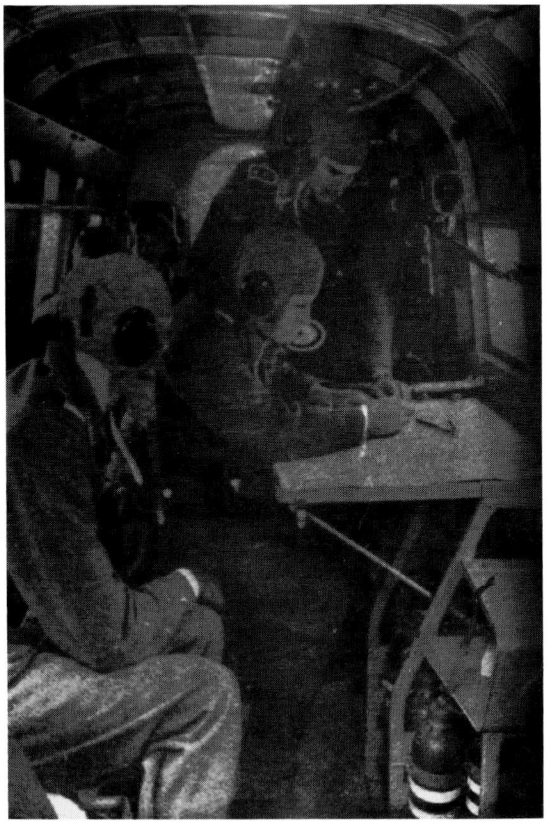

Left: Navigational training in a flying classroom.

Right: Detail photo of the nose of a Siebel Si 204 D-1 showing the insignia of Flugzeugführerschule FFS (B) at Aigen Enstal.

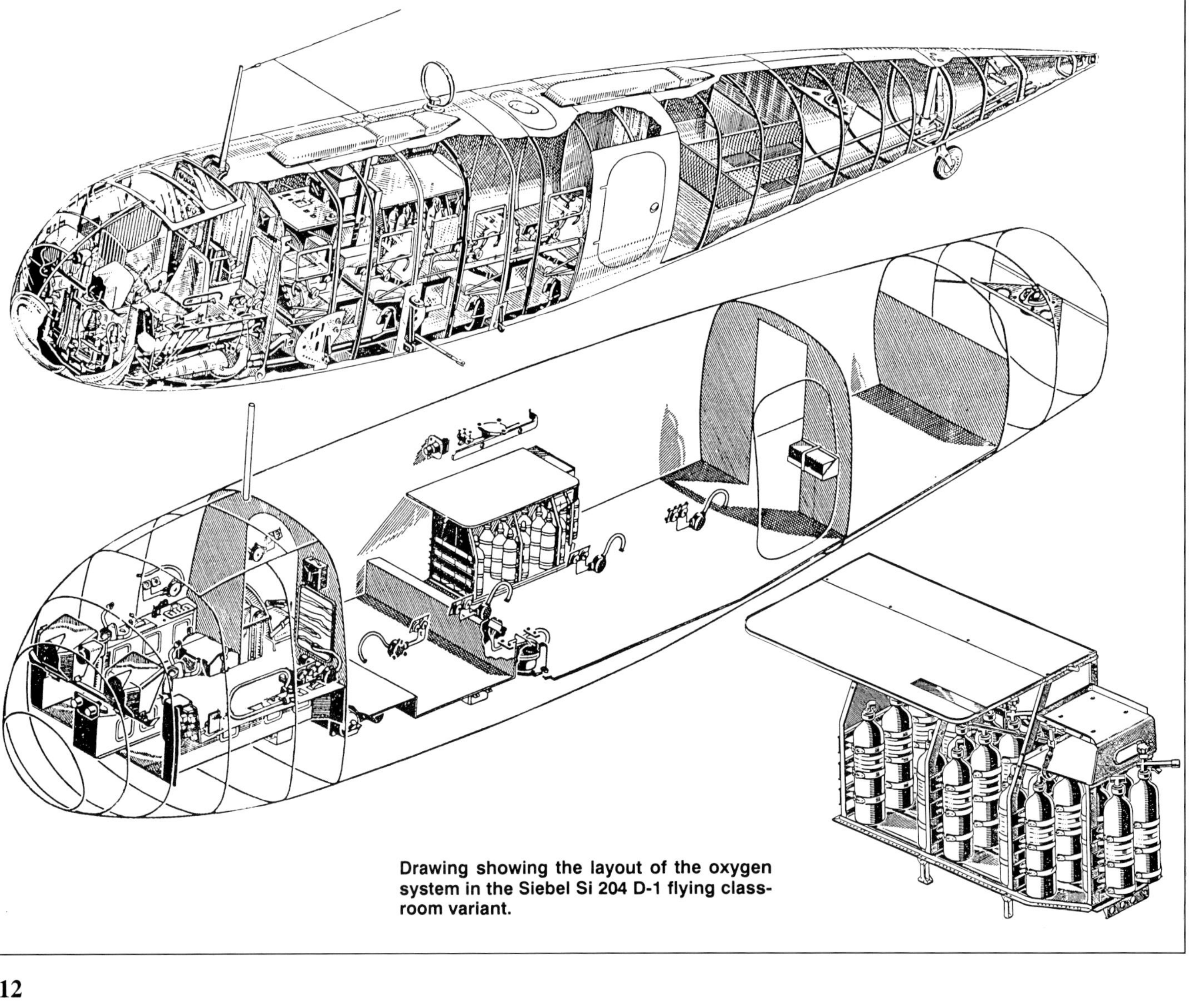

Drawing showing the layout of the oxygen system in the Siebel Si 204 D-1 flying classroom variant.

Siebel Si 204 D
Landing Gear Design

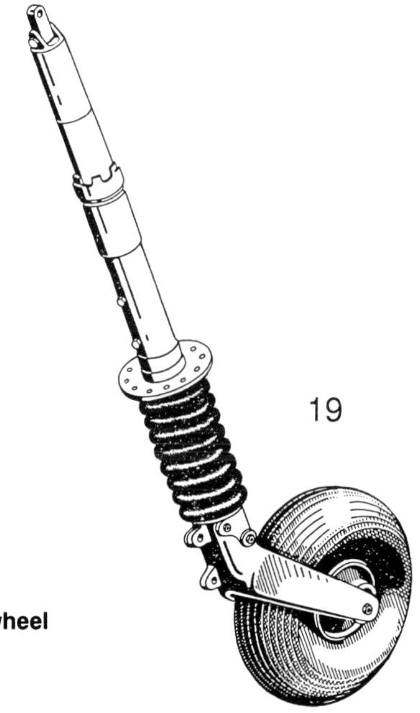

19

Tailwheel

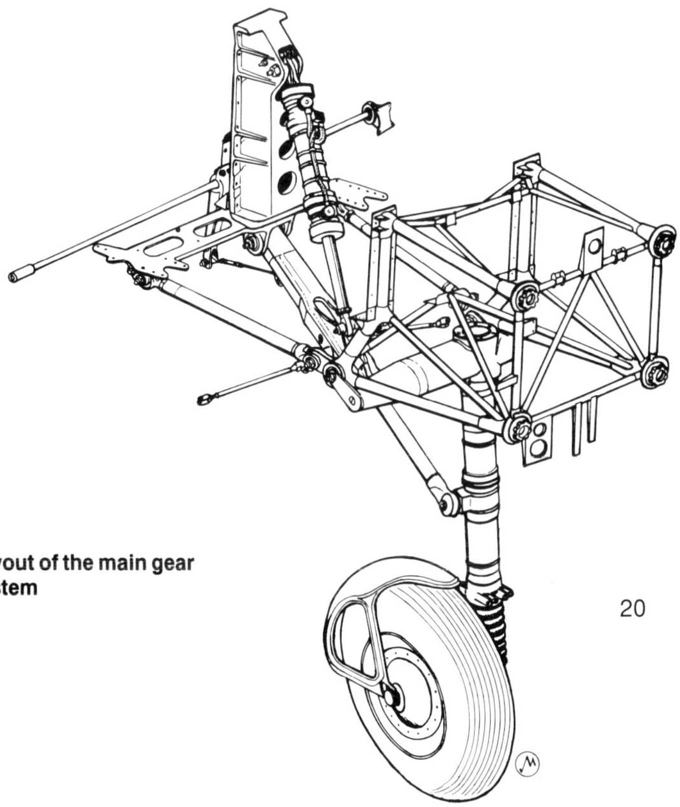

20

Layout of the main gear system

German Siebel Variants

Siebel Si 204 A

Type: Courier and passenger aircraft, generally same as prototypes, but with greater fuel capacity and electrically controlled variable-pitch Argus propellers.

Engine: 2 x Argus As 410 A-1 each with 600 hp.

Siebel Si 204 D-1

Type: Training and liaison aircraft. Fully glazed forward fuselage. Hydraulically controlled variable-pitch Argus propellers, self-sealing fuel tanks, Patin long-range compass system with automatic DF, landing lights.

Equipment: Kärcher heating system for cockpit and nose sections (H-1). Additional fuel tanks in outer wings and liquid de-icing system for propellers (B-1/B-2).

Siebel Si 204 D-3

Type: Training aircraft. As D-1, but with modified fuel system. Basis for French and Czech variants.

Technical Data (D-0/D-1)

Length:	12.01 m
Wingspan:	21.28 m
Wing area:	46.00^{2}
Max. takeoff weight:	5600 kg
Max. speed:	330 km/h (1000 m)
Service ceiling:	7000 m
Range:	1690 km

View of the Siebel Si 204 A's cockpit from Flugzeugführerschule FFS (C) 9 in Pretzsch on the Elbe. The photo was taken during a B2 overland navigation flight on 9 July 1943. Pilot Kühn, copilot Schöpplein, radio operator Kruse.

Right: A Siebel Si 204 A during an engine run-up with pilot Helmut Kühn at the controls. It's interesting to note that the Argus automatic pitch control for the propeller is not rotating while in idle mode.

Below: A Siebel Si 204 D of Flugzeugführerschule FFS (C) 9 at Pretzsch/Elbe during a formation flight. Its code is TJ+XG.

Primary components of the struts, shocks, tires and landing gear covers.

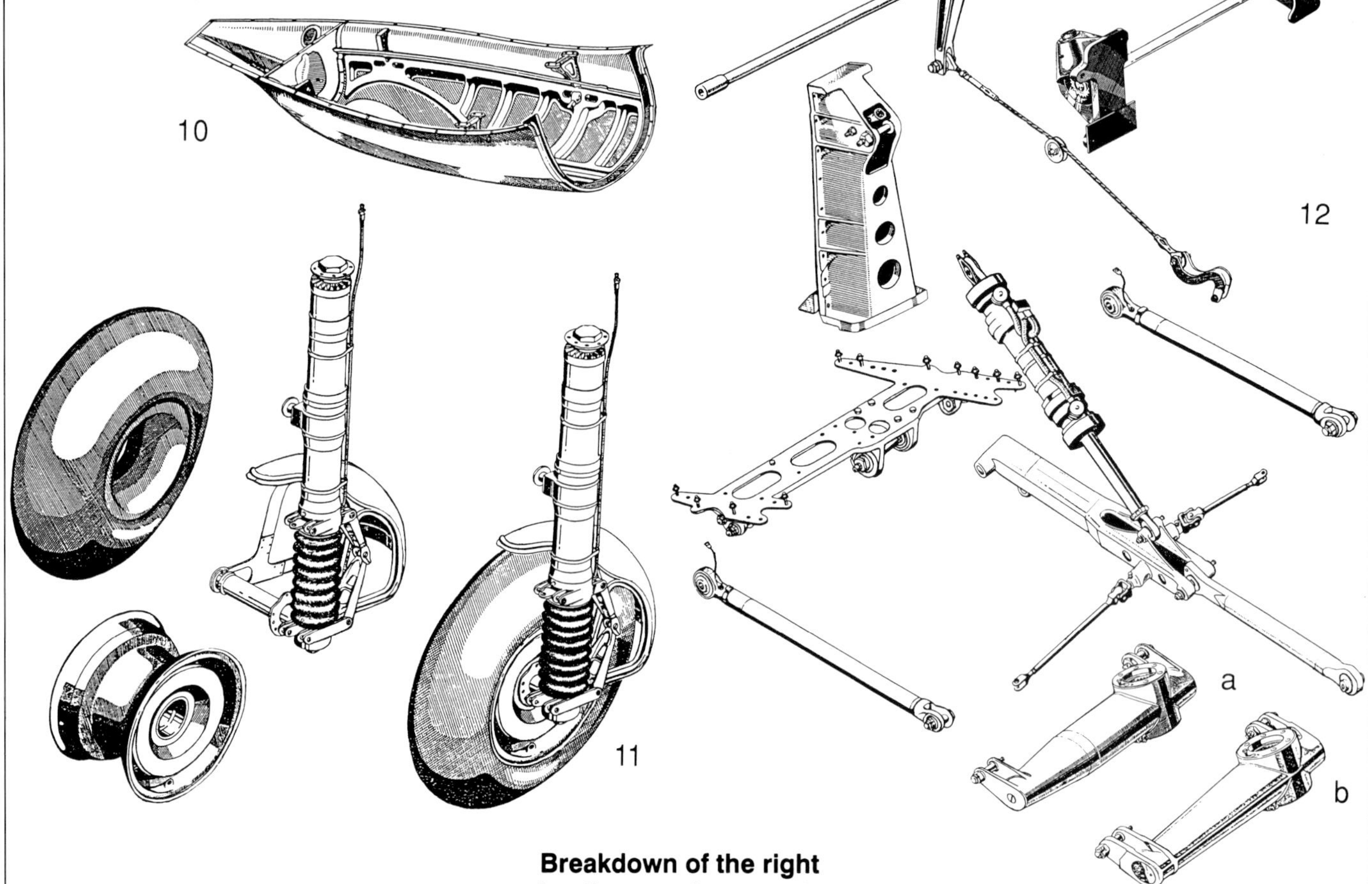

Breakdown of the right landing gear's retraction and locking mechanism.

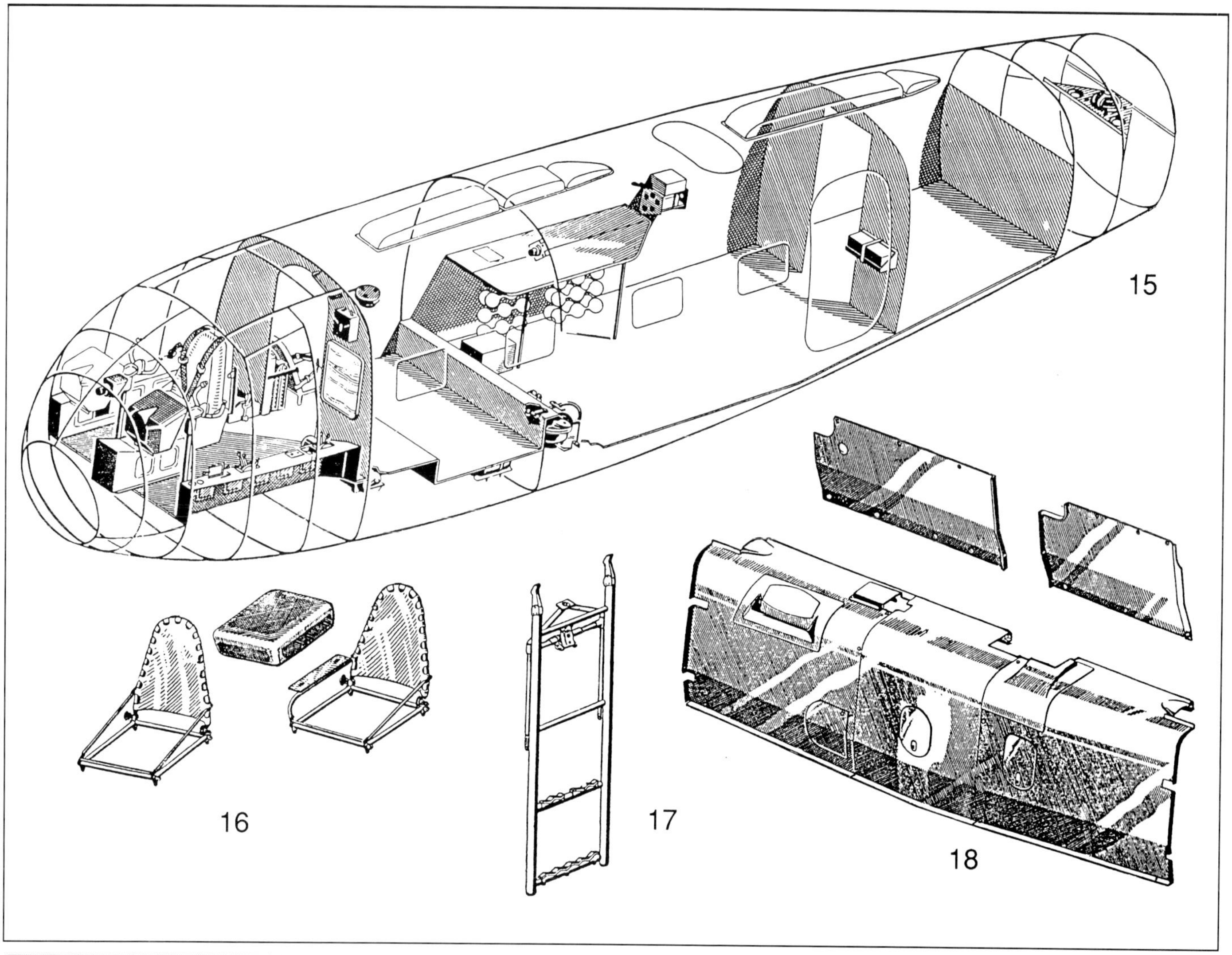

Detail drawings from the Siebel Si 204 D handbook.

Formation flight training with three Siebel Si 204 Ds from Flugzeugführerschule FFS (B) 11 at Aigen-Enstal in 1941, with the majestic Alps providing the panoramic background.

The prototype of the Siebel Si 204 E with a fully loaded weapons suite.

The Siebel Si 204 E as Bomber

The exact number of Siebel Si 204 Es built or converted cannot be determined with any accuracy from previously published material. The fact is that two prototypes (V 22 and V 23) were modified as bombers.

The high load carrying capacity of the Si 204 D (1000 kg) made the aircraft particularly suitable as a weapons platform, although its slow speeds virtually made daylight operations impossible.

This line of thought was pursued further in Czechoslovakia, where the armed version of the Si 204 served the Czech Air Force as an interim primary bomber until the arrival of the jet bomber generation. The events leading to the creation of this variant are described later in the "AERO C-3" section within this volume.

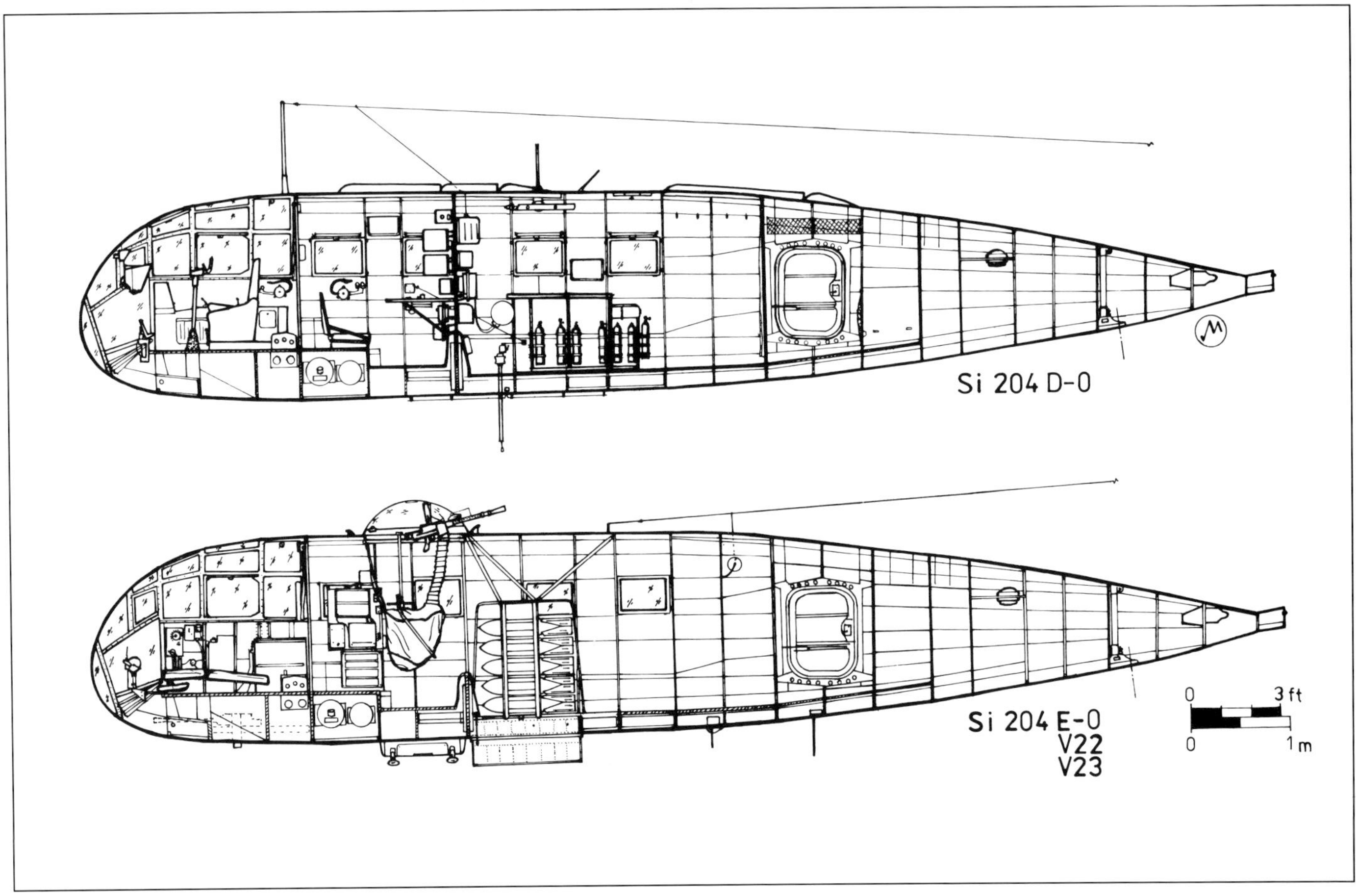

Left column (top to bottom):
1. Nose with MG 131 muzzle on left side and instrument flying shroud. Notice the bombsight as well.
2. MG 131 folded down.
3. Control stick with aileron trim control knob, bomb release knob and firing switch for the fixed forward firing MG 131.
4. Ammunition box for the right MG 131.

Right column (top to bottom)
1. Supplemental armor for the pilot's seat with Revi 16a sight attached.
2. Ammo feed for the left MG 131 attached to bulkhead 7.

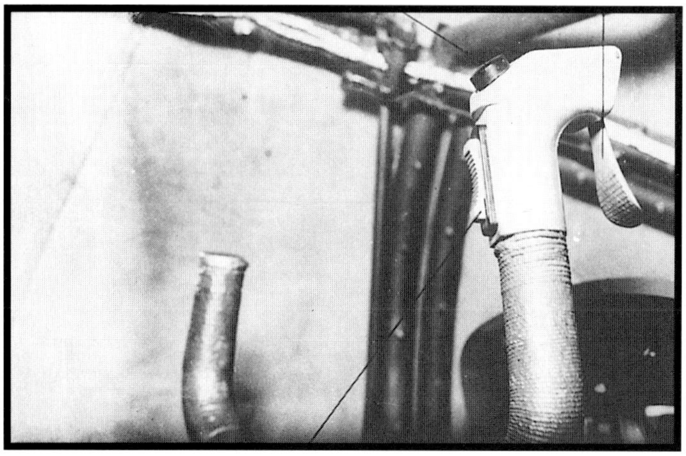

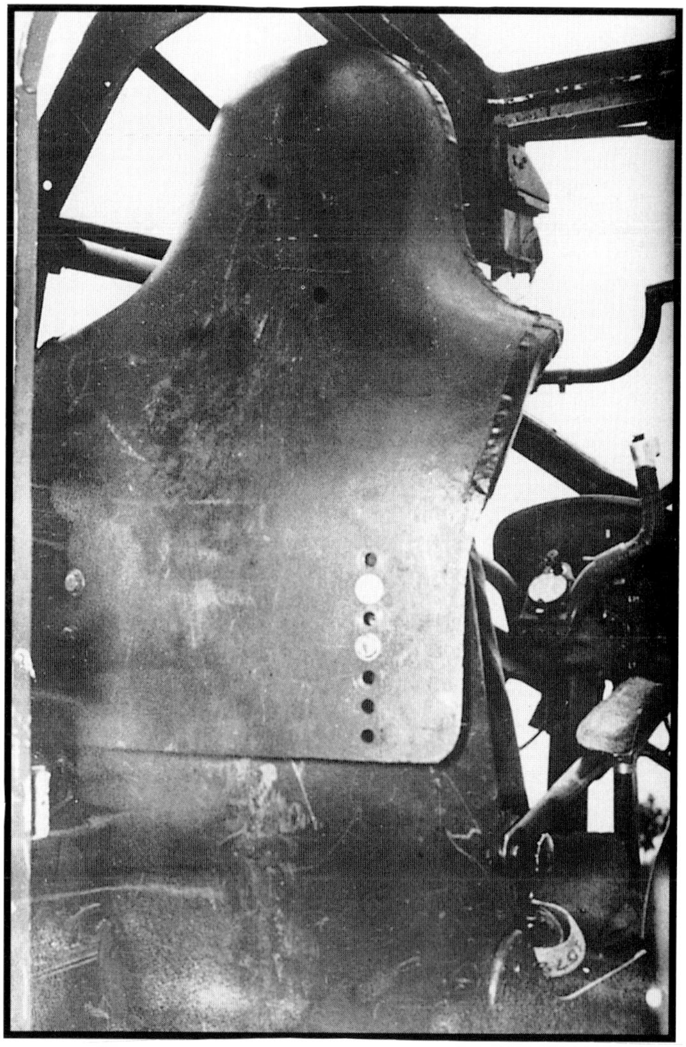

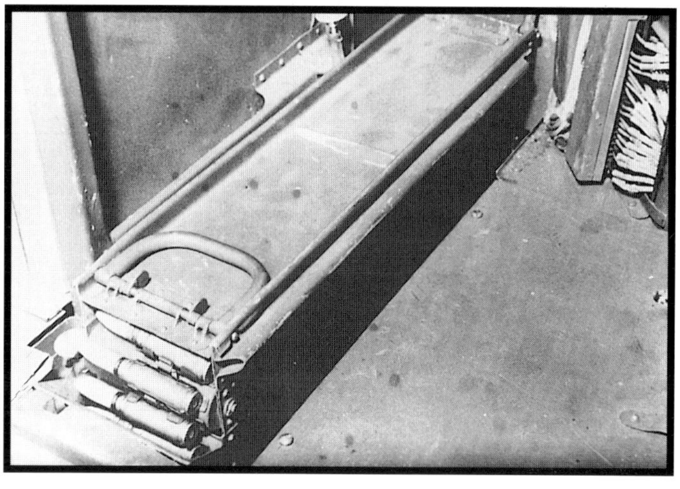

Detail photos of the bomb release system showing all technical components such as the intervalometer, emergency jettison lever, bombsight and electrical system.

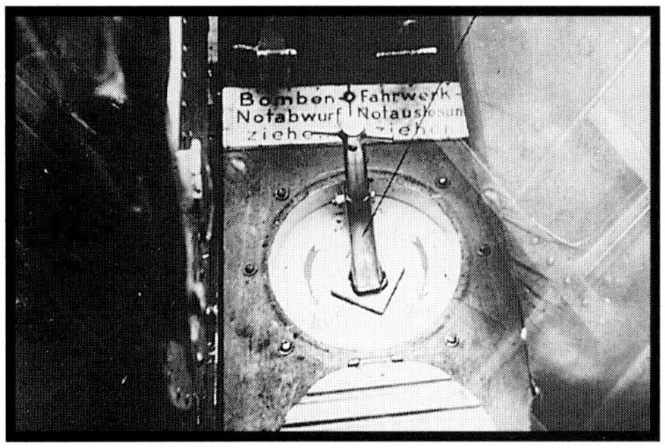

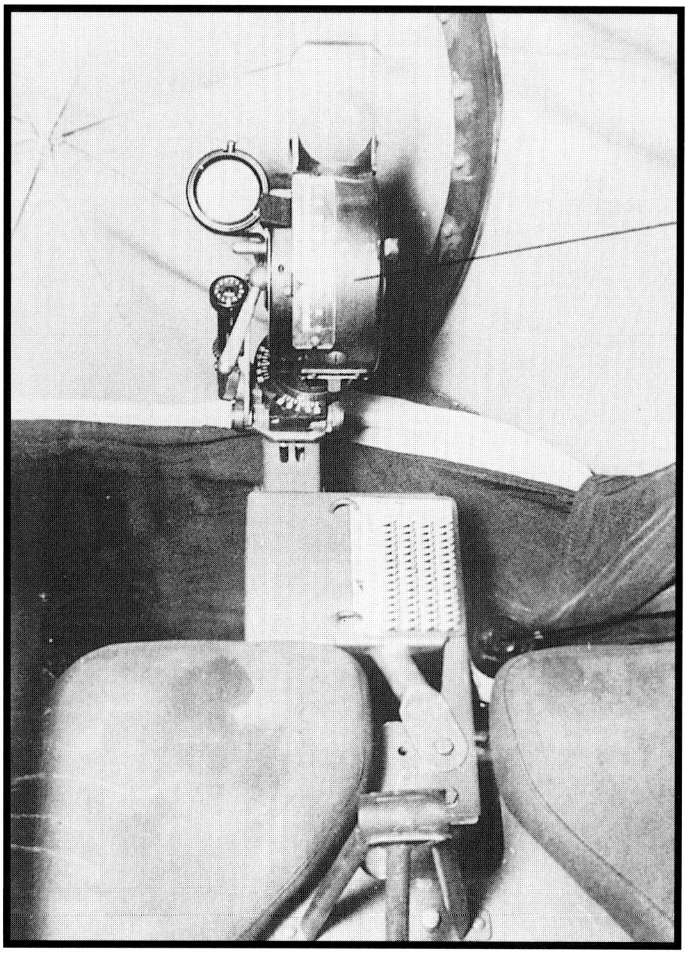

Left column (top to bottom)
1. Bomb bay doors seen from the inside.
2. ETC 500 bomb racks.

Right column (top to bottom)
1. Bomb bay door locking mechanism at bulkhead 10.
2. Bomb bay with 2 x 5 RAB 14 racks.
3. Open bomb bay doors from the outside.

MG 131 turret of the Siebel Si 204 E in detail

Left column (top to bottom)
1. Electrical turret mechanism with control knobs (on left)
2. Cartridge bag

Right column (top to bottom)
1. Within the turret, which could rotate through a full 360°, the MG 131 had a horizontal field of fire of 18° and could be moved vertically to +85°/-10°.
2. The MG 131 was installed off-center.
3. The gunner was protected by an additional armor plate behind the Revi 16a sight.

NC 701 of the Escadron de liaison aerienne (ELA) in the early '50s.

Siebel Si 204 in French Service

Siebel Si 204 production in France
Because the Siebel Flugzeugwerke Halle GmbH was operating at capacity with the production of Ju 88 and Ju 188 aircraft, Siebel Si 204 production was, in addition to Czechoslovakia, also moved to France.

The Companie Societe Nationale de Construction Aeronautique du Centre was selected to produce Siebel Si 204 Ds according to Luftwaffe specifications, and an initial order of 450 aircraft was placed.

However, only around 150 of these were built by the time the Allies liberated France.

Production got underway only with difficulty, on the one hand due to a lack of material and qualified personnel and on the other hand because of ongoing sabotage activities.

After liberation it was the stated goal of the De Gaulle government to arm the French military anew. Included in this rearmament program was naturally the air Force.

In order to promote the expansion of the new aviation industry and to minimize the development risk inherent with new designs, priority was given to continued production of proven German designs.

Thus we find that the Junkers Ju 52/3m, Focke-Wulf Fw 190 and Siebel Si 204 D were put back into series production.

As early as spring of 1944 captured Si 204s were transferred the Forces Aerienne de l'Atlantique, a partisan air force operated by communists. These were flown against German-occupied harbors and cities along the coast.

In France, series production was carried out under the following designations.

NC 700
Final assembly of Siebel Si 204 parts from German manufacture. 21 aircraft - with Argus As 411 engines - were built in this manner.

The airplanes were used by the French Air Force and navy.

Like others, the NC 700 was later re-engined with the Renault 12S motor.

NC 701 Martinet I
Identical to the Siebel Si 204 D-1 but with Renault 12S engines (license-built versions of the Argus As 411-TA) and three-blade propellers.

Prototype introduced in April of 1945, with construction of additional variants beginning in July of 1946.

NC 701 Version I
Air Force trainer.

NC 701 Version II
Naval trainer.

NC 701 Version III
Light liaison aircraft with seating for 8 passengers and luggage.

NC 701 Version IV
Liaison and command post aircraft.

NC 701 Version V
Navigational trainer.

NC 701 Version VI
Reconnaissance aircraft (first flight 6 January 1948)

Other modifications were known to have been carried out in addition to the above listed official types.

As with the Czech models, conversions were carried out in France for radio operator training, photogrammetry (four aircraft for the National Geographic Institute) and colonial patrol aircraft, armed with six bomb racks on external pylons (alternatively, four fixed and two traversable machine guns could be fitted instead).

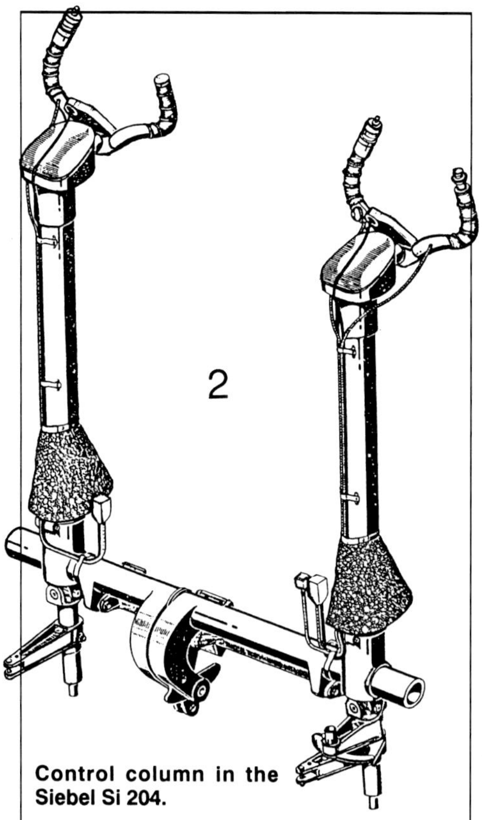

Control column in the Siebel Si 204.

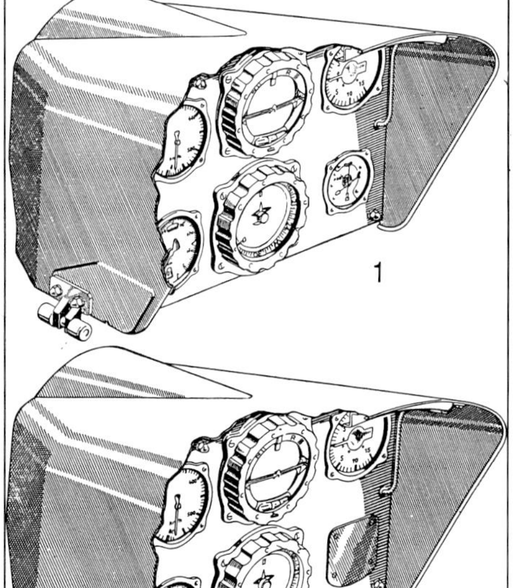

Left (top) and right instrument panels.

Maintenance work on a Renault 12 S.

NC 702 Martinet II

At the end of 1946 the SNCAC began production of the NC 702 Martinet II.

The most noticeable characteristic was the new nose design, which resembled that of the Siebel Si 204 A. Overall length increased to 12.81 m, with net weight at 3985 kg.

Altogether a total of 144 of this variant was produced. In 1949 production ceased at SNCAC, but was picked up and continued by SNCAN.

Martinets were used by the French Air Force in a variety of roles and saw service in nearly all theaters of operation.

Patrol flights were carried out in Africa and Madagascar.

The aircraft carrier Arromanches ferried a handful of aircraft to Indochina in November of 1948. Three squadrons used the type in Vietnam.

The last reconnaissance flight by a Martinet was flown over Dien Bien Phu in 1954.

Following the end of the French Vietnam War the airplanes were used in Algeria, where along with the MD 315 Flamant they were employed in launching the wire-guided AS 11 weapon.

As time went on the Martinet found itself being replaced by the more modern Flamant in active service.

The last Martinet retired from the Armee de l'Air in 1964.

A number of civilian airlines and businesses also used the Martinet, such as ONERA, SNECMA, AIR FRANCE, SAT, CATA, STAR, AIR ATLAS, SNCAC and SNCAN.

NC 701 in Polish service

The People's Republic of Poland bought seven NC 701s in 1946 for photogrammetry purposes.

From May 1947 the LOT Air Lines operated these aircraft.

On 15 May 1948 they were subordinated to the air force.

These aircraft provided the first comprehensive mapping of Polish territory after the end of World War II.

By the mid '50s, however, these machines had been mustered out.

Martinets in Sweden

Two of these airplanes were acquired by the National Geographic Institute of Sweden and used for photographic flights.

In December of 1964 three additional Martinets were also obtained.

All aircraft were former Martinets from the French Air Force.

Martinets in Morocco

Two NC 702 Martinets were delivered to Morocco's air force in 1960. However, nothing further is known regarding their role, service life or subsequent disposition.

Siebel Si 204 D of the Royal Canadian Navy as a coastal patrol aircraft. Unfortunately, no additional information is available for this example.

NC 701 Martinet I (N 6) of the French Air Force. The black and white stripes indicate that this aircraft may have been used in the Suez Crisis.

Siebel Si 204 E-0 light bomber

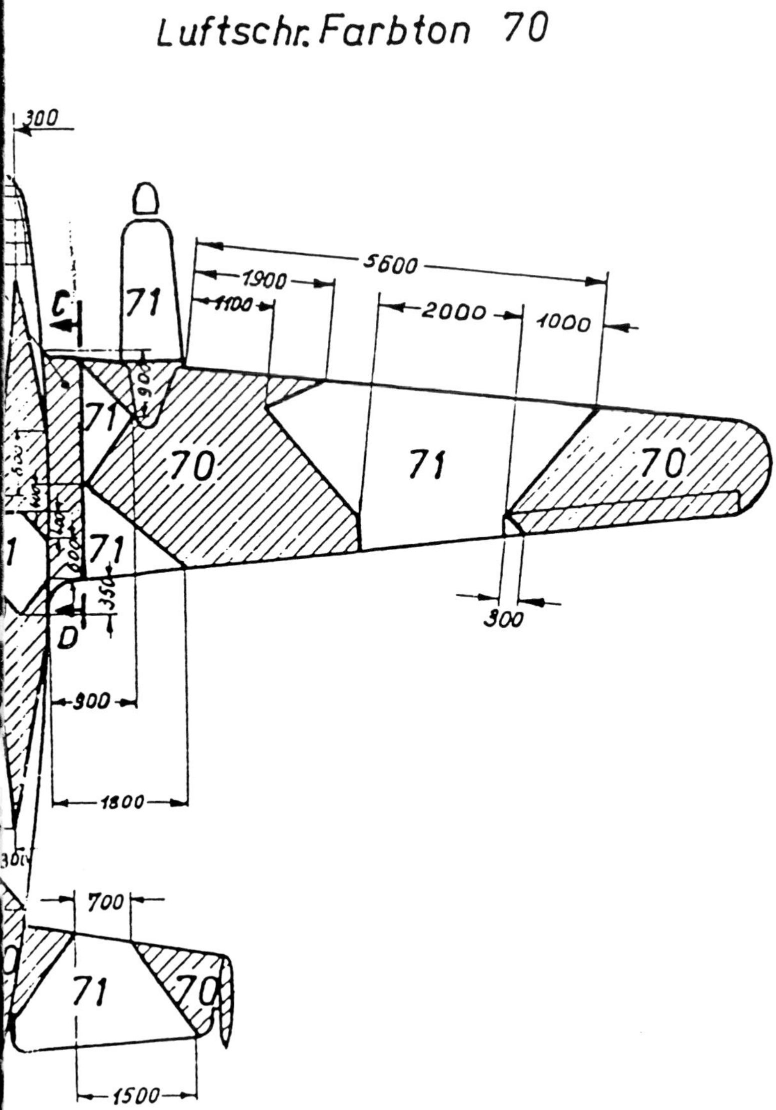

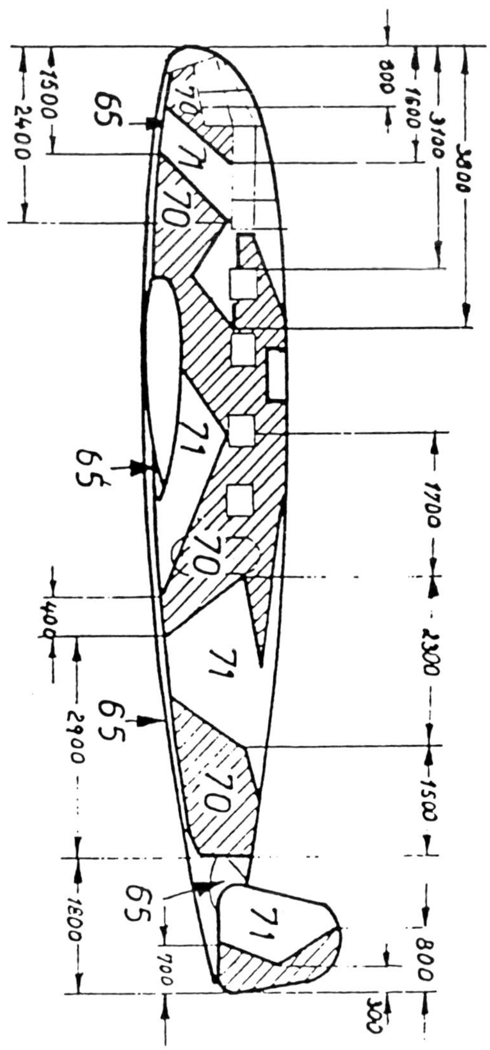

…ht 1:100

Der Reichsminister der Luftfahrt
und Oberbefehlshaber der Luftwaffe
Technisches Amt
Erprobungsstelle Travemünde

gez.: B o c k
2.2.42

Eing.: am. 15. 1. 4?
Lsch. 2667a

Änderungszustand	Oberflächenschutzliste zu.....	OS - Liste Nr...	Bl. Nr.
14	8-204.D-3 (vereinfacht)	8 OS 204.D-3	11

Colors

Upper surfaces:	**RLM 70 Schwarzgrün**
	RLM 71 Dunkelgrün
Lower surfaces:	**RLM 65 Hellblau**
Cockpit/wheel covers:	**RLM 66 Schwarzgrau**
Propeller:	**RLM 70 Schwarzgrün**
Spinner:	**RLM 71 Dunkelgrün**
Interior:	**RLM 02 Graugrün**

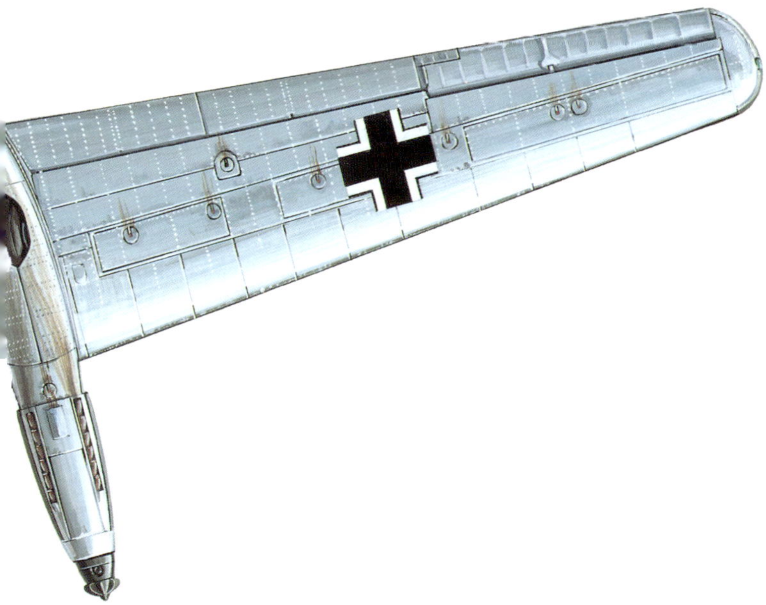

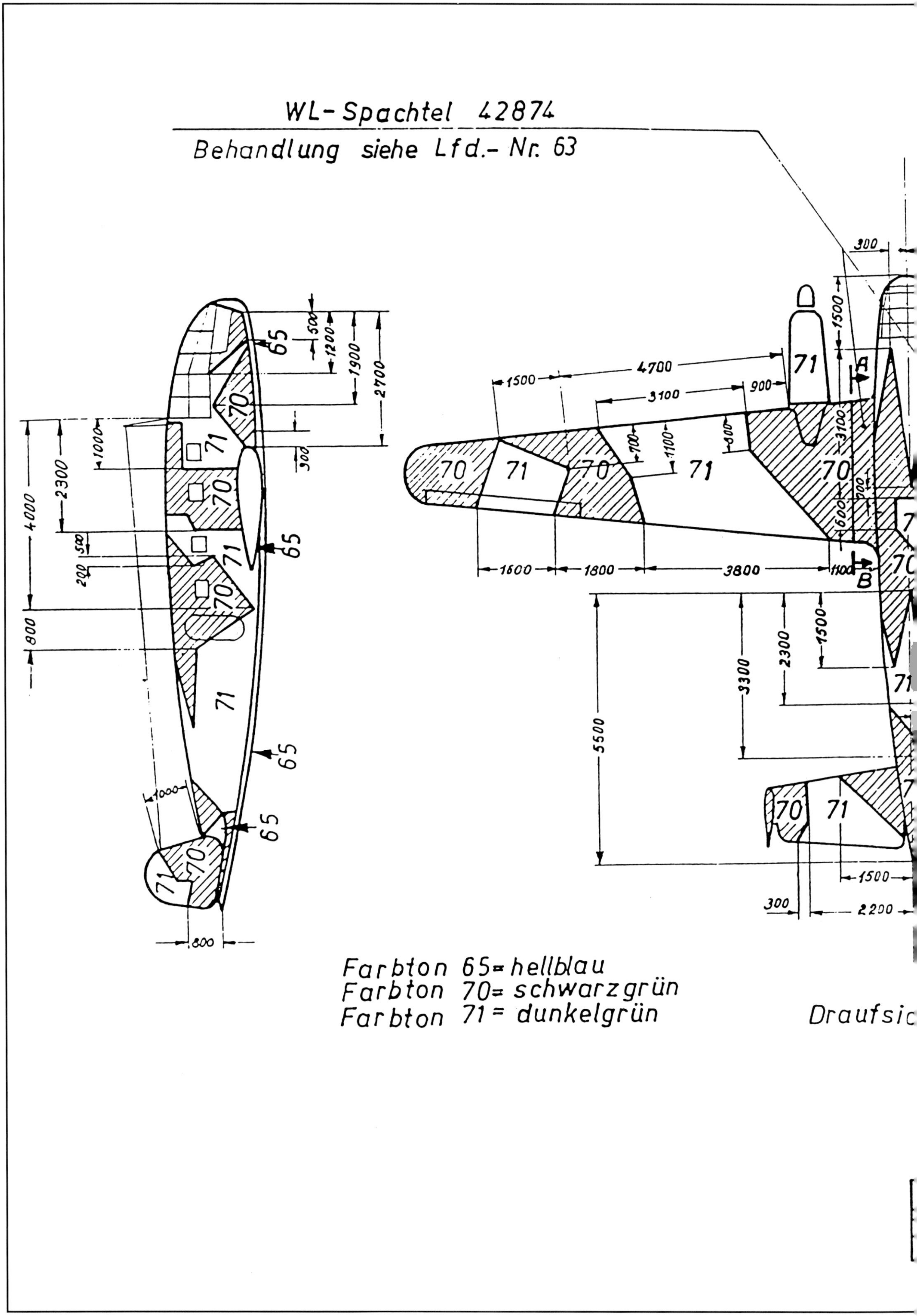
WL-Spachtel 42874
Behandlung siehe Lfd.-Nr. 63
Farbton 65 = hellblau
Farbton 70 = schwarzgrün
Farbton 71 = dunkelgrün
Draufsic

An NC 702 Martinet II serving as a testbed at a French test center. Notice the additional fin with pitot tube on the upper forward edge on the fuselage spine. The purpose of this testing is not known.

NC 702 Martinet II in civil markings. Registration is F-BBFR.

Two views of an NC 701 in a badly weathered state. Apparently demilitarized, the machine carries the civilian code F-BAOQ. It is not known whether former military aircraft saw continued service in any great numbers in civilian roles.

NC 701 Martinet I of the Aeronavale (French Navy). This photo was taken in 1953 at an unknown location.

This photo of an NC 701 Martinet I was taken in 1961.

Belly landing of a C-3B after gear failure. The splintered wooden propeller blades are clearly visible. Also notice the capped off upper turret. On-board equipment was generally similar to the Siebel Si 204 E.

Siebel Si 204/C-3B in Czech Service

In Czechoslovakia, production of the Siebel Si 204 was entrusted to the Aero Vysocany Works (LETOV) in Prague. Repair work fell under the responsibility of the aircraft repair facility at Prague Letnany.

Output of Siebel Si 204s between June of 1942 and the end of 1944 amounted to the impressive total of 1007 aircraft.

This reliable airplane's production and career was in no way over with the end of the war.

Series production began anew and continued under the designation of C-103/C-3.

A total of 179 aircraft were manufactured from 1945 to 1950.

Along with confiscated aircraft from former German stockpiles, these machines were used in a wide variety of civilian and military applications.

Civil aircraft received the designations of C-103 A, B and C, whereas military aircraft operated as the C-3A and B.

C-103 variants later flew operationally as army transports and were redesignated as C-103D or D-44.

Postwar production in Czechoslovakia was carried out on five basic variants.

C-103 C
First postwar variant, production of which was limited to just two aircraft.

Operated from June of 1945 in passenger and cargo transport roles.

Crew consisted of pilot, copilot and radio operator, plus seven passengers. Dual controls. Rudimentary seating facilities.

C-103 B
Improved passenger and transport variant, with seating for six passengers. Radio operator in separate compartment. Upholstered seating. Dual controls.

This C-103 A was used by the Svazarm sports sky divers. A safety cover was attached to the balance under the left elevator.

C-103 A
Passenger and cargo transport version for nine passengers. Upholstered seating with cargo hold. Single controls, radio operator in copilot's seat.

C-3A
Basic military version for IFR training.

Four crew members + four passengers.

Rudimentary seating. Some with cockpit curtain for IFR training.

C-3AN
Version for navigator and radio operator training. Cabin with navigation fold-up table. Astrodome on dorsal fuselage section.

C-3AF
Reconnaissance variant. Single controls. Crew consisted of pilot and three camera operators.

Three cameras normally carried: Rb 20/30, Rb 50/30, Rb 75/30.

Electrically operated lens covering.

Teardrop shaped fairing on the right fuselage side for installation of machine gun.

C-3AFG
Photogrammetry and signals services reconnaissance variant.

In use from 1950 to 1958.

Single controls. Crew comprised pilot/navigator and camera operator.

Camera configuration same as C-3AF, supplemented by 4 side and oblique cameras.

Teardrop shaped fairing on fuselage was for machine gun installation.

C-3B
Standard bomber variant. Replaced the retired Pe-2s and De Haviland Mosquitos until arrival of the new Il-28 jet bomber.

Equipment varied greatly among the individual aircraft. A single standard equipment suite cannot be ascertained.

The same aircraft from page 32 seen from a different angle.

Additional side windows along right fuselage with machine gun position.

For the most part, these aircraft were from indigenous production types. However, some were confiscated former German aircraft converted to this variant.

Bomb load in the fuselage and on external racks was up to 800 kg. (50, 100, 250 kg bombs)

C-3B aircraft gunner training variant
Fitted with four 13 mm machine guns (MG 131), each with 500 rounds.

Two machine guns were permanently installed under the fuselage, operated from the pilot's seat.

One machine gun in the right fuselage side window, plus one machine gun in the dorsal turret (DL 131)

C-3BV
Target tow aircraft for artillery training. Cable winch system installed in fuselage with release mechanism.

While in Czech service the C-103/C-3 variants proved themselves to be true jacks of all trades.

Additional roles included calibration flights, spraying operations for forestry, topographical flights and even bombing missions against ice-thaw areas along the Danube.

Main bases for military operations were Trencin (multi-engine training), Pardubice (navigator training) and Chrudim (radio operator training).

In addition, virtually all units within the Czech Air Force possessed at least one example as a liaison aircraft.

A C 103 of the Czechoslovakian Air Force after a crash landing at Budaörs, Hungary.

Above: 12-liter Argus AS 411 A with an output of 600 hp (445 kw). After the Second World War the engine continued to be built by the French as the Renault 12 S.

Left: Nose of a C-3AFG with additional teardrop side port and a yaw sensor.

Below: Loading a C-3B with bombs. The Revi 3b gunsight is clearly visible in the cockpit.

Siebel SI 204s Under Foreign Banners

This unique photograph shows one of the seven NC 701 Martinet Is acquired by the Polish Air Force in 1946 and used in the photomapping role.

A captured Siebel Si 204 D-3 operated by the Royal Air Force. The machine was thoroughly evaluated at Farnborough. The insignia of Flugzeugführerschule FFS (C) 5 in Neubrandenburg can be seen on the fuselage. It is not known when and where this aircraft was captured. Unfortunately, this example did not survive the postwar period.

Maintenance work on a Siebel Si 204 D in Dutch service. The attractive color scheme is quite interesting. Coding is PH-NLL.

Siebel Si 204 D-1 of the Swiss Air Force following a general overhaul. This is the same aircraft shown on page 10 which had defected to Switzerland and - as already mentioned - was integrated into the Swiss Air Force after being given a general overhaul.

Above: Siebel Si 204 D of the Soviet Air Force in the Caucasus (Khorog) in 1946.

Right: Another Soviet Siebel Si 204 D, built in Czechoslovakia, at the N II GVF flight test center in 1945.

Below: Siebel Si 204 D of a Soviet Arctic expedition with the markings CCCP N-35.

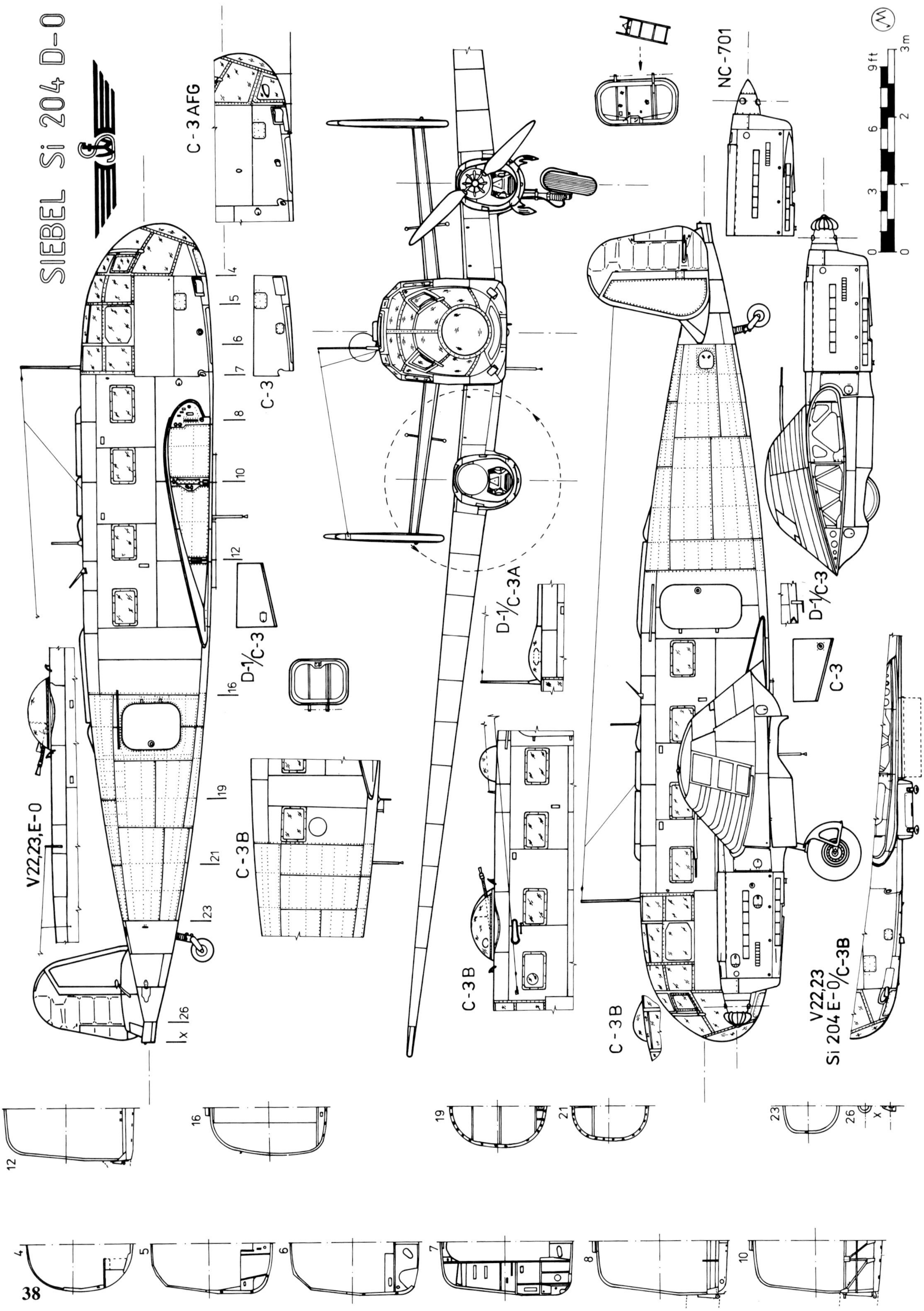
SIEBEL Si 204 D-0
C-3AFG
C-3
D-1/C-3
D-1/C-3A
C-3B
V22,23,E-0
V22,23
Si 204 E-0/C-3B
NC-701
9ft
3m

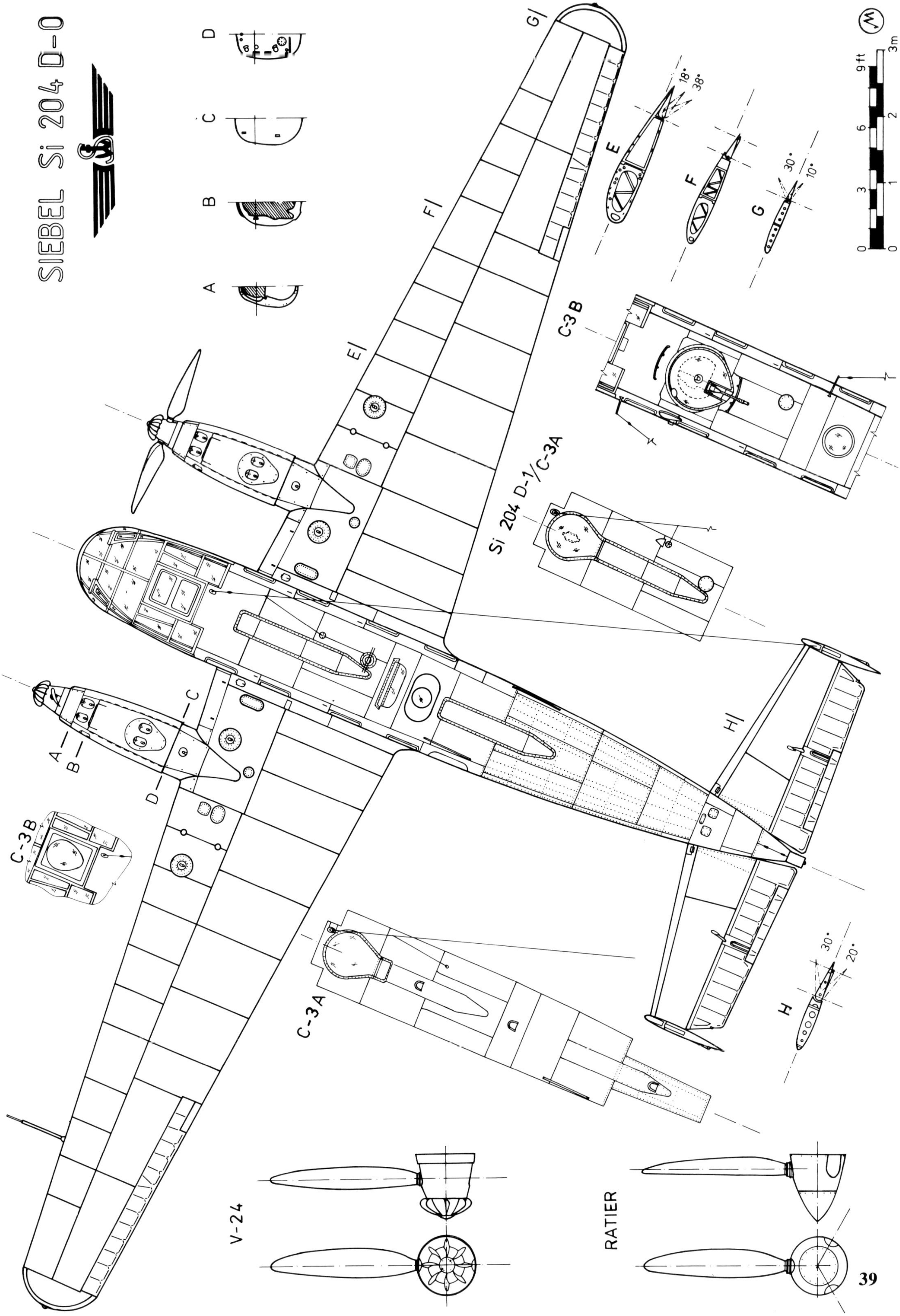
SIEBEL Si 204 D-0
A
B
C
D
E
F
G
H
18°
38°
30°
10°
30°
20°
0
3
6
9 ft
0
1
2
3m
C-3B
Si 204 D-1/C-3A
C-3A
C-3B
V-24
RATIER

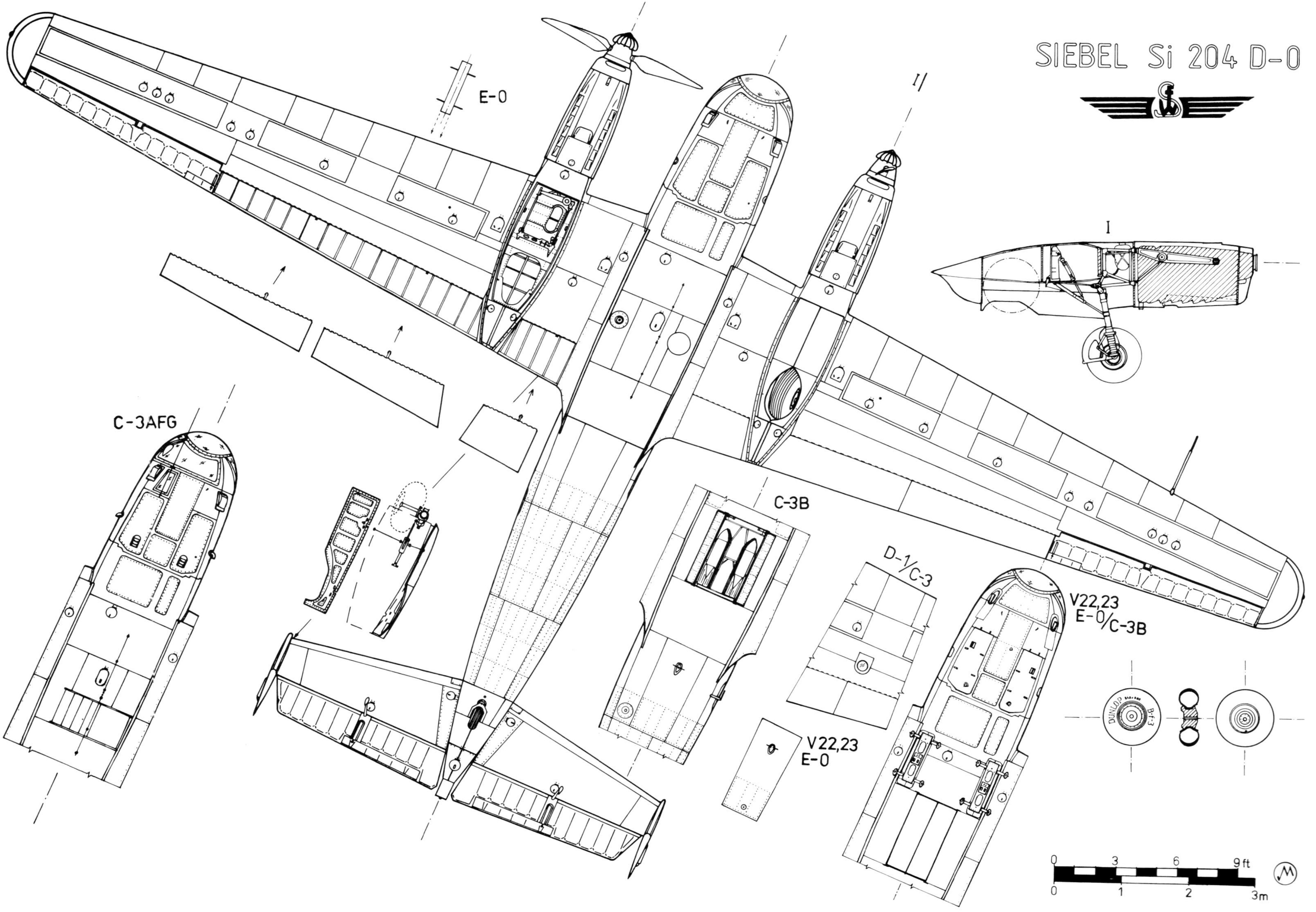
SIEBEL Si 204 D-0
E-0
I
II
C-3AFG
C-3B
D-1/C-3
V22,23
E-0/C-3B
V22,23
E-0
DUNLOP
B-f-3
0
3
6
9 ft
0
1
2
3m

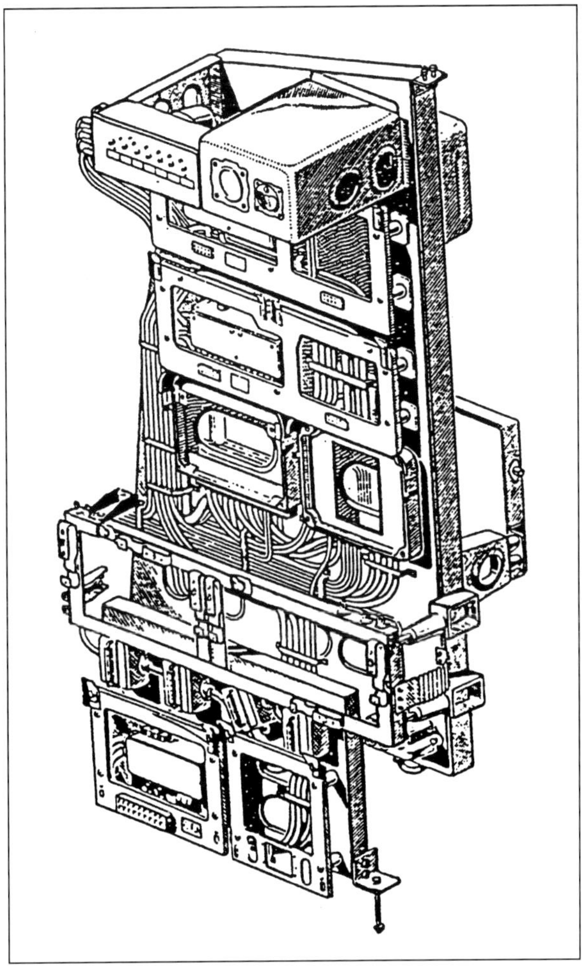

Radio equipment bulkhead minus radio components.

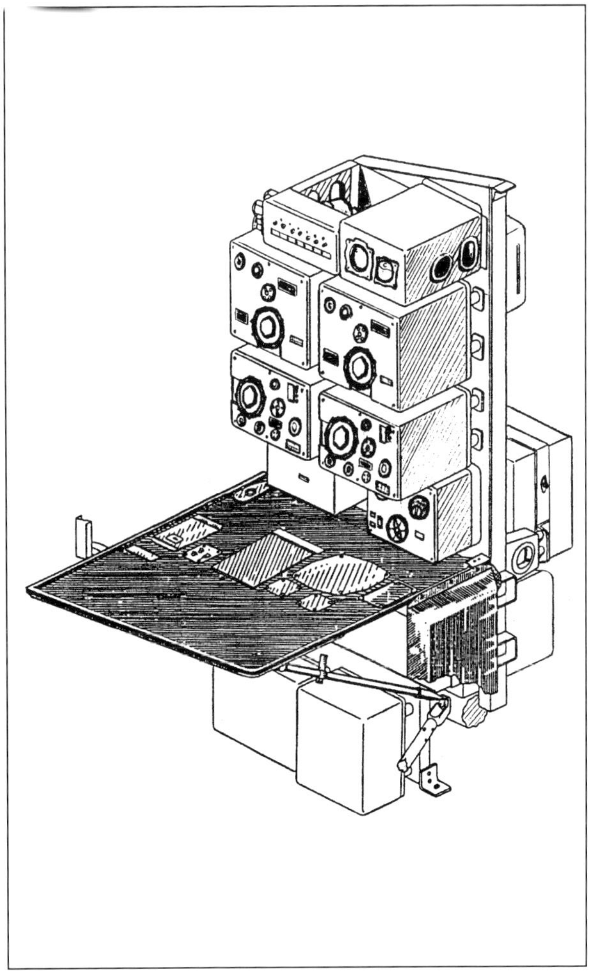

Complete radio equipment with navigator's table.

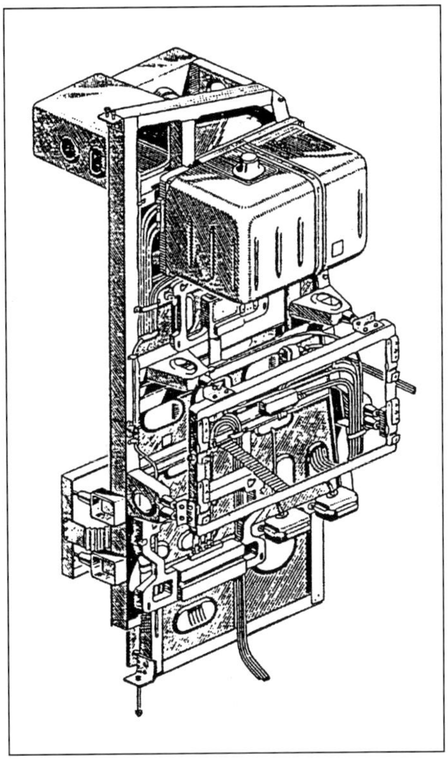

Rear view of the radio equipment support rack.

View of the radio operator's work station during flight.

The End of the Siebel Era

A Siebel Si 204 D parked at Ketterschwang in 1946. Unfortunately the insignia on the nose cannot be clearly identified.

Parked Siebel Si 204 Ds at a fighter training school. The picture was taken from the cockpit of another Siebel. Notice the sign at the edge of the woods (left edge of photo) which reads: "Feind hört mit" ("The enemy is listening").

In 1947 this Siebel Si 204 A was discovered at the edge of the Klagenfurt airfield. The insignia of the Flugzeugführerschule FFS (C) 11 at Zeltweg (Gemse) is prominent on the aircraft's nose. Markings were CJ+XM.

The same scrapyard at Klagenfurt seen from a different perspective. In the foreground is another Siebel Si 204 D with the code ?P+OB.

Above: This heavily damaged Siebel Si 204 D lay abandoned at the Neubiberg airfield after the war.

Right: Apparently used as a liaison aircraft, this Siebel Si 204 D was operated by the Americans. This photo was also taken at Neubiberg, although the date is unknown. It still flew in its original coding (SK+LF).

Below: A Siebel Si 204 D which crashed in April of 1945 was recovered from the Attersee in Austria in 1955 and was unfortunately scrapped.

Siebel Si 204 Carrier Trials with the DM 1

Another interesting role was reserved for the versatile Siebel Si 204 - carrier experiments with the Lippisch DM 1.

This Lippisch design was an unusual concept, discovered by American forces on 3 Ma 1945 at the Prien airfield near Chiemsee.

This radical layout by the legendary designer Alexander Lippisch was to have been a prototype for the Lippisch P 13a ramjet powered fighter. Its radical form was to have provided initial data in the realm of supersonic flight.

Since the DM 1 was unpowered, carrier trials were to have been conducted piggyback fashion using a Siebel Si 204, along with catapult launches.

Some sources indicate that a Siebel Si 204 A was to have been used. In other sources a Siebel Si 204 D is described. The drawing shown here comes from the book "Die geheimen Wunderwaffen des III.Reiches" by Mercado/Miranda. It shows a Siebel Si 204 D with the corresponding DM 1 piggyback configuration.

As no further detailed information is available regarding the carrier trials, configuration of the planned design remains speculative.

The next pages show a built-up model in order to better illustrate the concept.

This rare bird was discovered by American troops at Prien on 3 May 1945. The exact designation was Lippisch DM 1.

The unusual configuration of the DM 1 impressed the Americans so much that on 9 May 1945 arrangements were made - by General Patton personally - to continue building the DM 1 discovered in Prien.

For the Americans' part, consideration was given to using a Douglas C-47 for carrier testing, and apparently one of these aircraft was even reconfigured accordingly. Carrier trials never materialized, however, as wind tunnel testing took priority in the United States.

The DM 1's weight would have proven to be no problem for the Siebel Si 204's impressive load carrying capability. A greater problem was its slow speed.

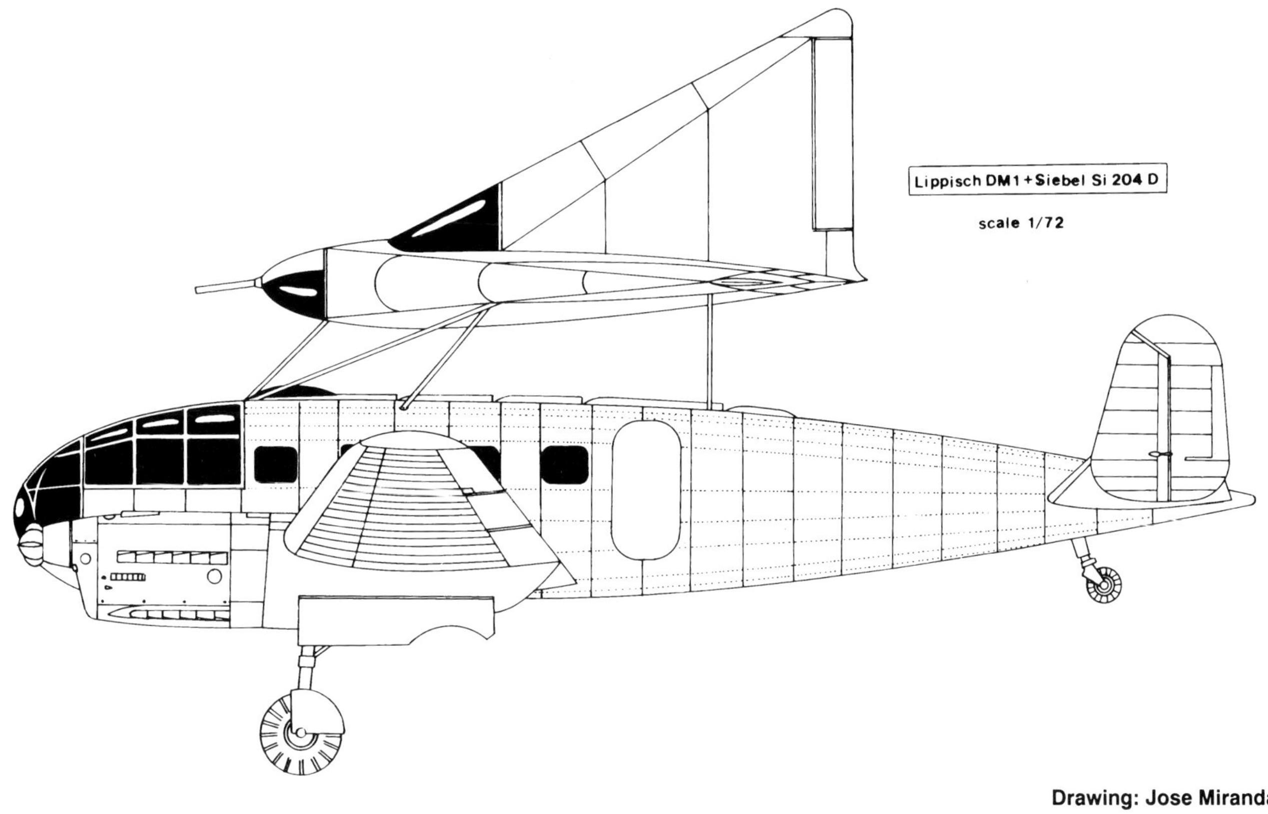

Drawing: Jose Miranda

The radical aerodynamic concept of the DM 1 necessitated a very high angle of attack during the landing approach. In order to improve the view for the pilot on landing the lower nose section was made of plexiglas so that the pilot could maintain visual contact with the ground while landing.

No other photograph could better emphasize the general layout of the DM 1, which was made entirely of wood. This is the state which the Americans found the aircraft in when they captured it on 3 May 1945.

Here rockets would probably have been attached to the frame supports in order to provide additional thrust.

Speeds up to 800 km/h were registered.

The concept was thoroughly evaluated in wind tunnel testing in the United States. The results, however, were anything but satisfactory and testing didn't continue. Some of the features were incorporated into the Convair XF-92 experimental aircraft.

Unlike the many Siebel Si 204s the DM 1 was preserved in the basement of the Smithsonian Institution. Of the many Siebels built, only a fuselage section and a wing survive today in the Czech Air Force Museum in Prague/Kbely.

This is a tragedy, particularly in view of the fact that Siebels were flying up until the end of the '60s.

One can easily imagine how fascinated the American technical experts must have been when they first glimpsed this design, yet another wonderbird from the German Luftwaffe. In the background are wing spars for another DM 1 prototype.

Translated from the German by Don Cox

Printed in China.
ISBN: 0-7643-0566-2

This book was originally published under the title,
Profile: Siebel Fh 104/Si 204 Varianten
by Flugzeug Publications.

We are interested in hearing from authors with book ideas on related topics.

Schiffer Military/Aviation History
Atglen, PA

Published by Schiffer Publishing Ltd.
4880 Lower Valley Road
Atglen, PA 19310
Phone: (610) 593-1777
FAX: (610) 593-2002
E-mail: Schifferbk@aol.com
Please write for a free catalog.
This book may be purchased from the publisher.
Please include $3.95 postage.
Try your bookstore first.

This model conversion in 1:72nd scale provides a better illustration of the planned piggyback configuration. The Siebel Si 204 D comes from FLUGZEUG, while the DM 1 is a Huma model.

Due to its high load carrying capacity the DM 1's weight posed no problems for the Siebel Si 204 D. It is interesting to compare the extreme differences between the more old-fashioned Si 204 and the almost futuristic looking design of the DM 1.

Above: An NC 701 Martinet I in the colors of the Swedish Royal Family. No further details on this aircraft are known. It carries the code SE-KAE.

Right: The last known remnants of a Siebel Si 204 D is this fuselage section in the Czech Air Force Museum in Prague/Kbely.

These two rare advertising posters from the Siebel company show the most advanced commuter airplanes of their day, the Fh 104 Halore (left) and its successor, the Si 204 A (right).